Jeziel Silveira Silva

Educational Practices in Geography

Jeziel Silveira Silva

Educational Practices in Geography

Experiences from everyday school life

ScienciaScripts

Imprint

Any brand names and product names mentioned in this book are subject to trademark, brand or patent protection and are trademarks or registered trademarks of their respective holders. The use of brand names, product names, common names, trade names, product descriptions etc. even without a particular marking in this work is in no way to be construed to mean that such names may be regarded as unrestricted in respect of trademark and brand protection legislation and could thus be used by anyone.

Cover image: Provided by the author

This book is a translation from the original published under ISBN 978-613-9-67119-9.

Publisher:
Sciencia Scripts
is a trademark of
Dodo Books Indian Ocean Ltd. and OmniScriptum S.R.L publishing group

120 High Road, East Finchley, London, N2 9ED, United Kingdom
Str. Armeneasca 28/1, office 1, Chisinau MD-2012, Republic of Moldova, Europe
Printed at: see last page
ISBN: 978-620-7-68607-0

INDICE

In recent years, the use of technology applied to education has become one of the main goals and objects envisioned by public education policies. In this field, schools have started to invest heavily in technological devices in the school environment, with the aim of improving the quality of teaching and learning. The creation and implementation of computer laboratories, the availability of resources (audio, video, TV), multimedia projectors, has been gaining momentum among teaching spaces, directing teachers to adapt and use these devices more and more in their classes, creating new educational methods and means, overcoming the barrier of traditional education: blackboard and textbook.

It seems that there are countless unattractive lessons and tasks that help students to lose their enthusiasm for lessons and their willingness to study, and that there is no connection between the knowledge and experience that students already have. In this way, students don't understand why they are studying the subject and the result is that they just memorize without understanding the topics.

It's quite common to hear students complain about geography content in schools. In particular, about the teachers' lessons, claiming that they are unattractive or uninteresting, presenting boring and unimportant content, as well as a lack of motivation and cooperation on the part of the students. CAVALCANTI (2010, p.3) points out that

Geography teachers are constantly fighting against barriers to make their lessons attractive to students "because most of them aren't interested in the content that this subject deals with".

By analogy, it is necessary to list two fundamental factors that help the teacher in the context of winning over the students: the first encompasses mastery of the content, which will make it less necessary to use the textbook; and the teacher's enthusiasm, the way he or she delivers the lesson (expressing him or herself in a clear and orderly manner), the use of the blackboard, how he or she moves around the room and the differences in intonation, avoiding monotony.

It is therefore essential for the teacher to analyze the target audience to be reached and what resources are needed to achieve greater attention and a reduction in classes that are considered unattractive by the students.

Finally, the aim of this work is to offer teachers activities based on teaching resources that can be used in the classroom and school space. With this, the reader can use the experiences in this book to transpose them into the classroom and improve the teaching-learning process in a dynamic, playful and different way.

CHAPTER 1: SCHOOL SPACE AND THE DIFFICULTIES OF PRACTICAL LESSONS

The advance in the use and implementation of technological devices in Brazilian schools is still recent. When many teachers arrive at school, they are faced with a primitive environment, where only textbooks are used to propagate content or where they lack the basic infrastructure to teach. For this reason, a large number of teachers who find themselves in this situation end up leaving their workplace or settling for the structure on offer, thus making their lessons repetitive and boring.

In the first instance, in order for this situation not to become routine, the teacher must bear in mind that traditional teaching resources (blackboard or textbook) and technological resources (data show and videos) are indispensable tools in the teaching and learning process, but that they can be alternated with other resources that are easier to access.

Secondly, the list of teaching resources on offer to teachers without technology is vast. Newspapers, games, magazines and photographs are just a few examples of materials that teachers can use in their lessons to develop creative, educational and playful work.

The advance of differentiated educational practices in the classroom has been growing more and more to reduce this lack of technological apparatus and resources. On the other hand, designing a creative lesson should be part of every educator's curriculum, but unfortunately they feel insecure and lock themselves into a conservative attitude, in addition to other factors such as a heavy workload or a demotivating salary, which leads to the exclusion of a differentiated lesson, which ends up becoming just a plan in their class diary or applied in a class where there are few students.

Notably, we would like to highlight another problem that is relevant in this field: the lack of funding for the creation or development of teaching materials or tools. Although some teaching tools have a moderate value (such as models and photo murals), it is still possible to create pedagogical instruments for designing and creating a differentiated lesson using little money, which KAERCHER (2009) characterizes as zero-cost geography (gzc).

CHAPTER 2: TEACHING GEOGRAPHY THROUGH PLAY

As a result of my experiences in the classroom and the use of various activities (games, music, photographs, videos, slides), I thought about creating dynamics to be used in the classroom in the future. The teacher, as the main individual in the classroom, chooses to experiment (sometimes) with different resources in the classroom and, in the end, draws up a comparative table, listing the positive and negative aspects of the practices carried out.

In general terms, the geography teaching that has been applied in Brazil's primary school system does not meet the interests of the students in a positive way, which somehow ends up affecting the educator as well, who is trapped by not attracting the students in the right way. Therefore, there must be an articulation between teaching, teacher and student, inserted in an objective, added to methodologies (teaching components), for the learning of geography teaching.

As mentioned above, we are surrounded by problems in teaching, which makes it difficult to develop pedagogical practices. In this scenario, Zero-Cost Geography comes to the fore:

> [...] it doesn't involve any extra costs or technological resources (nothing against them, but they are generally not available in public schools). A simple photocopy and we often already have the raw material for beautiful discussions and productions. The difference isn't the computer, it's making the class "click" (KAERCHER, 2009, p.135).

Because of the countless difficulties they face at work, there are still those teachers who base their work on the unalterable pleasure of spreading meaningful learning of the content they teach, including their students in broader professional projects. However, despite all this commitment, it is still possible to find unmotivated students, which contributes to the exclusion of creative and playful lessons.

At the moment, the inclusion of playful activities in the classroom has become a great partner for teachers, who can achieve satisfactory teaching-learning results through these activities. In this sense, the use of play as a didactic resource stands out because:

> Teachers interested in promoting change can find in the playful proposal an important mechanism for making geography teaching meaningful and, consequently, meaningful to reality. This can help to reduce the high failure and dropout rates found in schools, as well as stimulating students' interest in geography classes (SILVA, BERTAZO 2013, p.28).

In addition, the use of playful activities aimed at teaching geography provides pleasure and fun during lessons. In this sense, their increase in classes helps to develop and expand cognitive and motor skills in students (attention, perception, reflection), as well as becoming essential for the construction of knowledge that contributes to the formation of individuals, as stated by PIAGET (1975), since:

> [...] games and playful activities become significant as the child develops, with the free manipulation of varied materials, he starts to reconstitute and reinvent things, which already requires a more complete adaptation. This adaptation is only possible when the child itself evolves internally, transforming these playful activities, which are the concrete of its life, into written language, which is the abstract (PIAGET, 1975, p. 156).

Teachers are responsible for planning and implementing activities that develop the individual's structure. These activities take on a significant and challenging role, becoming capable of amplifying the knowledge already acquired by students based on their reality.

Finally, the choice of content and proposed methodologies must develop the student's capacity, provoking problematic and challenging situations, as well as stimulating creativity based on the proposed lesson content (SILVA, BERTAZO, 2013).

CHAPTER 3: TEACHING RESOURCES AND THEIR APPLICABILITY IN GEOGRAPHY TEACHING

The production and creation of teaching materials by teachers arises from observations and situations in the classroom. As a way of motivating and facilitating the process of teaching and learning content in the classroom, production presents itself as a significant tool, "as well as 'emancipating' the teacher from being a 'mere consumer' to being a producer of knowledge" (SANTOS, 2014). In this sense of emancipation, Castellar (1999) emphasizes that:

> Teachers must therefore act to appropriate their experience and the knowledge they have in order to invest in their emancipation and professional development, acting effectively in curriculum development and ceasing to be mere consumers (CASTELLAR, 1999, p. 52).

However, it should be pointed out that the creation, production and use of teaching materials in themselves do not disqualify a content-based lesson, because the material alone is not enough to teach. There needs to be a symmetry between theoretical, didactic and methodological knowledge and the material used.

In the midst of the innumerable means of information, some resources for improving the teaching and learning process are neglected. The use of printed information can make a significant contribution to the student, especially when these resources are attractive enough to develop differentiated teaching materials.

Due to the various situations I have witnessed during my academic career in supervised internships and PIBID, I will describe some teaching resources that can be applied to the classroom and school space, using low-cost artifacts or those present in the student's life.

3.1- FILM

During the observations made at the elementary school during internships I and II, it was possible to notice the lack of support materials for the classrooms, such as an overhead projector, an environment room and even an appropriate space for physical activities. As an analysis of my time at the school, I'm going to present the proposal for a project at the school: CINE ESCOLA. There is nothing new about using audiovisual resources in the classroom, as CAMPOS (2006) points out below:

> **There is nothing new about using audiovisual resources as a teaching aid**. You can use music, slides, photos, poetry, literature and films as illustrations and to better understand the content. It is always an instrument for learning. Cinema, as an art, has the advantage of being able to use the various forms of language used by the other arts, thus managing to communicate with depth and involvement. As with any art, cinema expresses, directly or indirectly, the values of the author of the script, the director, society and the historical moment in which it was made (CAMPOS, 2006, p.1, emphasis added).

The project called: ***CINE ESCOLA***, aims to use a Geography class at least once every 15 days for its operation. However, due to the lack of infrastructure at the public state school, a partnership can be established between the school's teachers, for example, if the school is fragmented in terms of its school space, Physical Education classes can be used for this situation. In observation, students spend most of their lessons inside the classroom, surrounded by games or content that is sometimes not considered attractive or repetitive. Therefore, in view of this, a partnership between Geography and Physical Education classes would serve as a support for students to develop skills in both areas, mediated by films, as well as introducing new models of production to the teaching-learning system, as ARAUJO (2007) indicates below:

> The inclusion of new ways of constructing the teaching-learning process is a necessary measure for an integral education that is appropriate to the cultural characteristics of citizens in modern societies. Cinema becomes an

3.2- Which movies should I use?

The choice of films must be made carefully, and the teacher must analyze a number of aspects, including: what I want to convey and what objectives I want to achieve with this film. In my context, I'm going to report on my experience with the films I've used in class.

The city of Sao Joao del Rei is very rich in culture. There are countless cultural approaches present in the city, including architectural constructions from past centuries, houses that were home to important personalities in times gone by, the widespread presence of gastronomy festivals, handicrafts, and other elements. At the same time, the city is inhabited by a number of high-profile citizens who contribute their knowledge and skills to the city's culture and local knowledge. In the first part of the project, the choice of films that refer to some historical-cultural-geographical context of the city of Sao Joao Del Rei, with the aim of highlighting the cultural importance of the city and bringing individuals closer to past and present times, passing on their knowledge, teachings and values through documentaries, which become relevant to the school environment, as discussed by BRITO et al (2011):

Introducing cinema into the classroom can become a way of exploring a variety of themes from students' everyday school lives. In **this way, cinematographic art provides students with the formation of critical and reflective minds with a considerable degree of understanding and interpretation of reality.** Communication is essential in contemporary interaction processes, so cinema in this context collaborates with the schools' pedagogical project (BRITO, FREIRE et al, 2011, p.1, emphasis added).

The use of films in the classroom is growing all the time. From this point of view, cinema can be most appropriate in the classroom in the form of documentaries, films broken up into two or three lessons or short fiction films. The short films have a duration adjusted to the classroom time, which can be accompanied by debates, conversation circles, workshops, among other things, as CAMPOS (2006) points out:

> **Cinema can be most useful in the classroom in the form of documentaries or short films. These make it possible, after preparation, to play the movie and discuss it during the course of one lesson. It doesn't seem quite right to use two or three lessons, on different days,** to play a movie and only discuss it the following week. In this case, depending on the social class of the students, it is more important to advise them to watch the film at home for analysis in class (CAMPOS, 2006, p.2, emphasis added).

Here I present some documentaries made by filmmakers from the region, dealing with themes of extreme importance to the social, cultural and geographical context of the city, and their assimilation in the school space, which helps in the understanding of the subject and in the student's vision, as highlighted by SILVA (2009), since:

> The cinematographic vision, as an education, reinforces the educational perspective of discussions on controversial themes, with dimensions that bridge the gap between emotion and reason. **Through good films, teachers can connect life, culture, reality, fantasy and motivation.** Today, diversifying lessons is essential to advance our critical thinking (SILVA, 2009, p.2, emphasis added).

For this stage, it is suggested that the class be divided into small groups (a maximum of five students) and that each student be responsible for bringing a short film about the city to the classroom, thus developing an activity of research and particularities. During the supervised internship I & II, I noticed that the classes were more enthusiastic about geopolitics. In a similar vein, in this section I present the second part of the project, the aim of which is to show the students short films on the theme of the great wars (mainly: World War I[a]),

2ª World War and Cold War).
Figure 1: Short films about the city of Sao Joao Del Rei (MG)

Source: Youtube, adapted by the author (2018). Image A: Tiao Paineira; Image B: Picole do Amado; Image C: Everything I have comes from here; Image D: Sinoansias.

This relationship can be worked on in the classroom, relating it to certain aspects of the film, as CAMPOS (2006) discusses below:

> From a geographical point of view, perhaps some useful aspects can be raised for observation: the ideology of the author and director, the ethnocentric vision, the archetypes present in the picture, the authenticity of the landscapes and the framing options of the space represented. Normally, the places represented in the images are not authentic, the action does not take place in the place alluded to by the plot. Beautiful landscapes are built with the support of panoramic screens, paradisiacal places and forests are created in studio, without the marks produced by History (CAMPOS, 2006, p.4).

Below is a list of films from the Netflix online catalog that can be used in the classroom. Netflix is just one option for students, and its use in Brazil has grown a lot in the 21st century, as ROSSINI and RENNER (2015) explain:

> In Brazil, where the service became available in 2011, the number of subscribers reached 2.17 million in December 2014, according to an estimate made by Digital TV Research. With a more limited catalog due to restrictions on negotiating distribution licenses with producers, NetFlix offers only 127 national productions (ROSSINI, RENNER, 2015, p.4).

For the project, the best option would be for the students to watch the films at home and, during the course of the project, have small debates about the films they watched. It would be appropriate for each debate to include two films at a time, facilitating conversation and situations such as student participation and assimilation of the themes involved.

Figura 2: Temáticas Geopolíticas contidas no catálogo da Netflix

Source: Netflix, adapted by the author

For this stage, it is suggested that the teacher make a greater effort to carry it out, as he or she becomes the main one in the search for films that objectively and simply allude to the themes studied. In particular, it is also suggested to choose a movie for each classroom and discuss it over 2 or 3 lessons, thus providing a playful lesson, which can be transformed into a form of assessment, escaping the traditional methods used in the classroom.

When you have a project of this size in the classroom, whether in primary or higher education, we are presented with challenges and invited to analyze certain concepts in a different way. When you are the teacher, and you have this instrument to teach, in this case, the cinema, you open up a range of possibilities and teaching methods that will bring sensitivity, everyday facts and other sensations to the student or viewer.

In analysis, it is important to emphasize that through the creation of CINE ESCOLA, it is possible to share, learn and improve knowledge among students, teachers and others. Furthermore, the use of local films triggers another analysis. Often, we get hooked on something outside our place of origin. This is a normal part of being human, and the project makes it possible to recognize a little more about local cinema, to pay tribute to the directors and actors, and even to reconnect with the city's history.

In this way, I present CINE ESCOLA as something innovative, challenging and exciting to work on with these students, which has a limited environment and can be converted into a simple project using a language known to all: cinema.

3.3- GAMES

Games, associated with teaching practices, have become a valuable tool, thus playing a key role in mediating the content presented in the classroom.

The application of games is intended to break with traditional practices by triggering logical reasoning in the student, making the learning process more meaningful.

> [...] school geography has increasingly become an area of knowledge that is socially committed to the production of the human condition and the conscious production of spaces, be they natural, social, cultural or political. There have even been some initiatives on the part of educational systems which, even in a prescriptive way, have been drawing up curricular proposals guided by critical approaches which dialect, or at least seek to dialect, the discussions produced in the fields of both Physical and Human Geography (THIESEN, 2011, p.87).

In this context, I present the proposal for the creation and application of the game (called the "three clues game"), which was developed as a result of classroom observations during internship III, based on the need to arouse students' attention and interest in teaching and learning the subject, encouraging curiosity and awakening their desire to learn. Based on the school space, the application of the practice was mediated through games, during Geography classes, presenting a greater depth of theoretical content covered by the textbook, worked in the classroom, with adolescents, enrolled in the 1st and 2nd year, of High School.

> Using teaching resources to facilitate learning is of the utmost importance in any subject, but using these resources in Geography classes is even more important. The Geography teacher has the task of trying to get his students to relate in the best possible way to the space they inhabit and transform. However, this task is not easy, because they don't always have all the necessary resources at their disposal to be able to demonstrate to their students all the complexity we have in relation to both nature and society (FRANQA, 2009, p.4).

As a proposal, the dynamics should be carried out in groups, where each group will have the right to answer the colored envelopes. Each envelope contains a series of questions, which will be asked randomly to the group and can be divided between easy and difficult ("a matter of luck"). Each question will be asked using clues, where the number of correct answers varies according to the clue: 1^a clue = 100 points, 2^a clue = 50 points, 3a clue = 25. There will be 4 rounds for each group and the group that scores the most points wins the competition. The value of the activity will be analyzed with the teacher.

As a suggestion and as information, for the first year classes, we will use questions related to the Hydrology unit and for the second year classes, questions related to the world and Brazilian population. The estimated time for carrying out the activity is around two 50-minute lessons in each class. The groups can be separated during the previous lesson, and each group will need to read what has been worked on in class, together with the book.

> The **materials available to geography teachers are becoming increasingly varied and easily accessible.** When using teaching materials, teachers need to be aware of how they will be used and also be selective when organizing lessons. One of the resources that teachers use is different languages, as they are all responsible for the student's reading and writing skills and there is access to texts via newspapers, scientific journals and the internet" (CASTELLAR, VILHENA, 2010, p.65 emphasis added).

Another suggestion for games in the classroom or school grounds is to use games that make use of the school grounds, such as the playground, canteen and sports courts, if possible. Here are some ideas for games that can be used at school: Word Cover, Crossword Puzzle, Geography Bingo, Guessing, among others.

In this context, I would like to highlight a game I played in elementary school about orientation and location. The game was played by pupils in the eighth grade and was called a kind of treasure hunt. The teacher gathered small groups of students together and, using the satellite image of the school, spread out clues leading to the final treasure. For this, the use of orientation and localization were indispensable, since it was a question of clues and orientations. Each clue contained a direction, and at the end of the route, a box of sweets was found. The aim of the game was to work on questions of orientation in the geographical space and to bring Physical Geography closer to the students' daily lives, providing a greater understanding of the geographical space.

3.4- FIELD WORK

In this section, I'm going to report on an activity that comes from fieldwork. In the field of Geography, teachers find it very difficult to work with field activities and practical activities, such as laboratory experiments, due to a number of factors. As a practical and easily accessible solution, both in terms of location and funding, I report on an experiment in the context of Environmental Education.

Environmental Education encompasses various aspects of knowledge, presenting great complexity in its development, which may or may not awaken the student's awareness, causing possible environmental damage committed by them in the present and in the future. Environmental education is closely linked to the individual as a social being, so individual perception is worthy as a component of practicing or disseminating environmental education about the conceptions of each agent in the social space (DIAS, MARQUES, DIAS, 2016).

One of the many capacities of environmental education, soil education is seen as an educational process that favors a concept of sustainability in the relationship between man and the environment. Like environmental education, soil education is a formative process that needs to be dynamic, permanent and participatory, in the search for a "pedological conscience" and a sustainable environment (PERUSI, SENA, 2012).

In terms of importance, the space dedicated to soil in elementary school textbooks is often inert or relegated to a minor level, both in urban and rural areas. Furthermore, this content is often taught in a watertight way, without relating it to the practical or everyday usefulness of this information, causing disinterest on the part of both the student and the teacher. These parameters contribute to the population's ignorance of the importance and characteristics of soil, which increases its alteration and degradation (SOUZA, SILVA, OLIVEIRA, RODRIGUES, 2016).

This work aims to highlight the importance of playful lessons in the classroom, addressing the theme: The effect of burning on the soil in the process of water infiltration and the importance of vegetation cover as a mitigator of this process, preferably for elementary school students, with the aim of arousing interest in the subject and providing students with the necessary knowledge about the effects of burning on the soil, through differentiated classes and experimentation, thus providing the teaching-learning process significant, as praised by FERREIRA, RODIGUES, JESUS (2011), since:

> Teaching practice is of fundamental importance when it comes to working on content, as it helps the teacher to teach their classes, making them

Before entering a classroom, the mentor needs to understand what content will be taught, what method they will use with the class and whether their mechanisms are the most relevant to the level of the class (FERREIRA, RODRIGUES, JESUS, 2011).

In diagnosis, the practice presented in this work has the specific potential to investigate some important issues surrounding environmental education and soils, such as: The infiltration and retention capacity of water in exposed and covered soils; The importance of vegetation cover on the soil surface; Discussion of the environmental problems caused by fires; Establishing the concept of environmental awareness in students. In view of this, as a proposal for a playful lesson, an experiment was carried out using materials such as 2L plastic bottles (PET) and two 500 ml plastic bottles (PET).

Collecting soil is an important step in carrying out the practical lesson. As an analysis, it is important to separate an exposed soil, in the case of this experiment, one that has been burnt, from a clump of soil with vegetation cover. It is advisable that the soil with vegetation cover is created by the students, using seeds such as corn, sunflower and so on, planted in a patch of soil. It is not necessary to use high-value, sophisticated equipment to carry out these practices. In the absence of such equipment, it is possible, according to the reality of each school, for the teacher to make adaptations to his or her practical lessons using existing tools and also to resort to materials that are inexpensive and easy to acquire (CAPELETTO, 1992).

Figure 3: Soil experiment

Source: Author (2017). Image A: Material needed; Image B: Soil experiment.

By way of analogy, I would like to offer educators two other proposals for field lessons, now using the outdoors. Often, the school has extra funds to provide students with local trips and outings, where concepts can be worked on in a clearer way, moving away from traditional methods.

Here I present a proposal for fieldwork, reported in my internship IV, called: **The Italian Immigrant In Sao Joao Del Rei: Knowledge & Know-How.** The aim of the proposal is to show students a little of the Italian heritage in the city, visiting properties in the area and highlighting the importance of these people to the regional economy, their traditions and life stories. To do this, the methodology is to visit properties located between the settlements of Giarola and Felizardo, places that are home to Italian descendants. The journey will have to be made by bus, as the locations are around 10-15 km from the school. The points will follow the itinerary of "Agro turismo Colonia Viva", a project that already exists in the city, linked to tourism. Finally, based on the students' analysis and perceptions, they will be asked to give feedback on the fieldwork, a kind of report (no more than 10 sheets), in which they should describe their observations, putting them into context with the topic studied and other topics in Geography. Groups of 4 to 5 students will be formed.

Figure 4: Location of the field visit area, Sao Joao Del Rei (MG)

Source: Google Maps adapted by the author (2018).

The other fieldwork I'm highlighting is due to the dynamics that cities have been going through in recent years, which has led to numerous dysfunctions in the geographical space, including in the context of rivers. With this in mind, my proposal here is about the **Perception and Socio-environmental Impacts of Land Occupation in the stretch of the Corrego do Lenheiro and Rio Das Mortes (Sao Joao Del Rei, Minas Gerais).** The aim of the proposal is to show the students a little of the rivers that cut through the city and pass "by their side", bringing together issues discussed in class such as: river regimes, physiognomy, the issue of urbanized rivers, pollution and the use and exploitation of the river by human occupation. Thus, the work has 4 stops, located about 1-5KM from the school, being compatible with the journey on foot. Points 1 and 2 are analyzed from the perspective of Corrego do Lenheiro, which comes from the Tijuco district and cuts through the city's historic center. Points 3 and 4 are analyzed from the Rio da Mortes, which now begins to transport sediment from Corrego do Lenheiro and later meets the Rio Carandai in the north\northeast direction of the city. Finally, based on the students' analysis and perceptions, they will be asked to give feedback on the fieldwork, a kind of report (no more than 5 pages), in which they should describe their observations, putting them into context with the topic studied and other topics in Geography. Groups of 4 to 5 students will be formed.

Figure 5: Prepared field route

Source: Google Maps adapted by the author (2018).

CHAPTER 4: OTHER MATERIAL PROPOSALS

The production and creation of teaching materials by teachers arises from observations and situations in the classroom. As a way of motivating and facilitating the process of teaching and learning content in the classroom, production presents itself as a significant tool, "as well as 'emancipating' the teacher from being a 'mere consumer' to being a producer of knowledge" (SANTOS, 2014). In this sense of emancipation, Castellar (1999) emphasizes that:

> Teachers must therefore act to appropriate their experience and the knowledge they have in order to invest in their emancipation and professional development, acting effectively in curriculum development and ceasing to be mere consumers (CASTELLAR, 1999, p. 52).

However, it should be pointed out that the creation, production and use of teaching materials in themselves do not disqualify a content-based lesson, because the material alone is not enough to teach. There needs to be a symmetry between theoretical, didactic and methodological knowledge and the material used.

In the midst of the innumerable means of information, some resources for improving the teaching and learning process are neglected. The use of printed information can make a significant contribution to the student, especially when these resources are attractive enough to develop differentiated teaching materials.

Based on the practices provided by the Institutional Teaching Initiation Scholarship Program (PIBID), a workshop was held for 9th grade students. The workshop, entitled "The creation of a geographical magazine from images", was held for around 90 students, and its purpose was to create a magazine article (of an informative nature), created by the students themselves. Supported by a brief theoretical part on some examples of magazines, the differences between magazines, the use of Lead[1] and the basic elements for creating an article, the students were divided into groups of six, where they used cut-outs (in this case, engravings, images or photographs) to create an article with a free theme.

The materials had to refer to some branch of geography (cartography, agriculture, tourism, population, economics, geopolitics, among others). The plan used magazines and newspapers, scissors, colored pencils, writing pencils, white glue and rulers to make the initial material[2] . In terms of positioning, we highlight this fact because they are basic, everyday tools that don't require a lot of money and are therefore accessible to all age groups, fitting in with the GZC.

In prefiguring, we opted to use newspapers and magazines for clippings, given that:

> [...] it gives students contact with different languages such as texts, cartoons, strips, graphs, tables, images, etc. that allow them to contextualize geographical content (PAULA, TORRES, 2014, p.3).

In concept, the use of newspapers and magazines in the classroom, as a mechanism of communication, becomes significant in the pedagogical field insofar as the educator uses it in his practices as a method of teaching and learning, enabling the construction of students' knowledge and collaborating in the formation of creative and autonomous readers

(ANHUSSI,2009).

---

[1] It's a form used by journalists to create stories. These are the five questions the story has to answer: what, who, when, why, how and where.

[2] The materials were then digitized, created and printed in the form of a magazine, made available for collections by the author and by Clara Rita Rosa, a student on the Social Communication course at the Federal University of Sao Joao Del Rei, in 2017.

In this respect, Silva (2005) highlights the importance of using information from newspapers or magazines, because "it makes the lesson more dynamic and participatory, encouraging debate and the exchange of opinions, moving the student from the position of inactive recipient of new information to a more active subject in the process" (p.148).

The discourse that it is important to provide situations in the classroom where "students can express their ideas, opinions and, in short, their experiences, is heard by both thinkers and teachers" (PAULA, TORRES, 2014, p.6). Following this

thought, we are constantly bombarded with daily newspapers and weekly magazines, which become a valuable resource for the classroom, given that these resources establish a connection between the student, everyday life and current affairs. In relation to this, MALANSKI (2010) highlights this important connection, given that:

> The space in which students live is full of representations and meanings constructed from social perception, and bringing students closer to this reality values cultural factors and allows them to understand the different scales of geographical analysis. The newspaper can be a teaching resource to help in this process (MALANSKI, 2010, p.5).

In view of this, newspapers and magazines are sources of information that are connected to geographical themes, because the news brought by these media make social, economic, political and cultural references that occur in different geographical spaces, developing in students the perception of their lived space.

CHAPTER 5: BUT WHY CREATE A MAGAZINE INSTEAD OF USING READY-MADE ONES?

Creativity has become the focus of analysis by scholars, as it has been certified as an important aid to human development (GONQALVES, FLEITH, LIBORIO, 2011). However, what has been observed is that creative potential has been insufficiently stimulated in the school environment, thus triggering creative opportunities and limited stimuli in the classroom.

> [...] Creativity requires an important combination of factors linked to the individual, i.e. creativity depends on cognitive, conative, emotional and environmental factors, given that one component always acts in the presence of others (OLIVEIRA, ELENCAR, 2012, p. 542).

In recognition of this, ALENCAR AND FLEITH (2008, p.61) point out that there are some barriers that hinder the development of creativity, such as: "large numbers of pupils; pupils with learning difficulties in the classroom; low recognition of the teacher's work; pupils' lack of interest in the content being taught and the scarcity of teaching material available in the school; insecurity about trying out new teaching practices, lack of autonomy in the way teaching activities are conducted and lack of enthusiasm for teaching".

But are these barriers really unbreakable? Why not use Zero Cost Geography? Why not propose something different that attracts the student's attention? Couldn't the difficulties be eliminated or reduced through a playful lesson? Large numbers of students, but no group work?

As a comment, most teachers are not suited to change and end up resisting the idea of proposing something different in their classes, but the role of the teacher is crucial for change. Even when the administration is innovative, teacher resistance can hinder or prevent the implementation of a change project.

In this way, the absence of change does not create a space that develops the capacity for interaction, socialization, competition, relaxation and cooperation, thus making it resistant and fixed, unable to break away from the walls of the classroom and the traditional model: chalk, blackboard, textbook and speech.

In their pedagogical practices, schools are still reluctant to work with and interact with different cultural languages in their daily contexts. In this context of cultural languages, we would like to highlight another coefficient that is part of magazines and newspapers, which are presented as alternative languages and which are highlighted to develop creativity, but which are often approached in the classroom without the student's immediate reality, serving only as an illustration: visual culture.

In appreciation, visual culture is part of students' lives, even before they enter school. According to CAVALCANTI (2010), visual culture can be characterized as:

> [...] the set of images that run through their daily lives has to do with the dissemination of a constructed subjectivity of the world, which also contributes to the construction of the subjectivity of the beholder (CAVALCANTI, 2010, p.10).

In view of this, it can be seen that, despite the use of alternative language activities, these need to be incorporated more into everyday lessons and worked on in a way that is consistent with the content, mediating between knowledge and learning and not merely gesturing as illustrative parameters or distant from the student's reality.

When we look at how teaching resources are used in geography teaching, we see that there is a wide range of opportunities for educators to use teaching resources (BRANDAO, MELO, 2013). We can therefore see that the use of teaching resources has become indispensable when teachers want to adapt their lessons to a more dynamic and attractive profile. Well, teaching resources are listed as facilitators of learning, as they are consolidated as a connection between the content worked on in class and the student's learning.

However, we understand from our own experience that the production of teaching materials by the teacher is capable of achieving good results in student learning, but there are insufficient incentives for this practice, given that too many of us are reproducers of ready-made knowledge and do not initiate new teaching methodology tactics, making the search for transformation, progress and change even more difficult.

But we should point out that we are not against the teaching methodology used by professional educators. We are alluding here to the fact that new resources, whether existing or created, are capable of shaping classroom dynamics and providing a relaxed atmosphere, helping students to understand and realize that their reality is closer than they think

At the same time, we need to see that the educational process is an agglomeration of actions, attitudes, sentences, decisions and methods. This grouping is not without a certain degree of intentionality and systematization, which tries to transmute positions, principles, cultures, themes and pedagogical practices. In this interpretation, educational current affairs recognize heterogeneous panoramas, that is, various types of metamorphosis.

With regard to the supervised internship, I would like to point out that it is considered by many to be the first opportunity that undergraduate students have to get closer to school life, making them able to combine the theory learned in academia with the practice of teaching. It's worth pointing out that, even with different resources and classes, it's not always easy to apply everything that has been planned, because practice is a complicated and tortuous path, as it is subject to the social, cultural and operational dynamics of a given school community. What can be observed are the teachers, who with total satisfaction enter the school to fulfill their role: that of educator, even though there are discouragements caused by indiscipline, lack of professional recognition from the government. Another aspect that should be mentioned is the role of the family and the community, who, together with the school, are able to contribute to the student's education.

We can highlight two key factors in the rise of quality education: the first comes from the barriers to be overcome in basic schooling, which is due to the traditional teaching method, both on the part of the school and the teacher, thus improving teaching skills; and the second comes from the physical space of schools, which is largely destroyed, precarious or lacking basic elements (e.g. video rooms, laboratories).

The practices and proposals for practical lessons in this article are just suggestions for educators to use in their classes. Throughout my life, I have considered that teachers' creativity should be put into practice, because when we are directed to teaching institutes, we end up getting used to traditional methodologies and forget to use tools to provide greater flexibility in the teaching-learning process. As a suggestion, it would be interesting to plan each step of the activity, get to know the dynamics of the classroom and the people who make up the school space. In my experience, I have witnessed countless cases of students' lack of interest in some activities and even personal slips in the application of activities. As human beings, we are subject to success and mistakes and we must pay attention to the main details.

I have to confess that working on large, playful and well-designed activities requires a huge amount of time, which we teachers often don't have in the face of countless daily tasks. In addition, the presence of 30-40 students in a classroom gives us a certain instability and discomfort with activities, as we are always insecure about applying different activities to a large number of people and afraid of failing and not being able to complete what is proposed. In this case, as I mentioned, it becomes necessary to divide the class into small groups, as this facilitates the dynamics of the classroom and works on personal concepts with the students, such as group interaction, team leadership, collective work, among others.

Another aspect to keep in mind is to always ask students for suggestions on how to improve lessons and content. We should always get out of our comfort zone and try different activities, because when the teacher doesn't change their teaching methods and resources, they become a simple reproducer of the content, leaving aside the phases and barriers of teaching and learning and making the student simply absorb everything, without questioning or asking questions that may be on everyone's mind in the classroom.

So I want to mention that my role in writing this book is not to criticize teachers who use traditional methods or those who only use differentiated teaching resources. Each teacher, in his or her work, in his or her personal analysis and analysis of the space in which he or she is inserted, whether that space is the classroom, the school or society, has his or her own way of writing and teaching. This writing simply recounts experiences throughout primary school, and tools and resources to make lessons more dynamic, enjoyable and diverse.

Finally, teaching practice is a particular way of constructing technical and scientific knowledge that belongs to the teacher's experience and personal journey, so that they can reinterpret, intercede, create and produce tools capable of providing educational support in the teaching and learning scenario (JUNIOR, NASCIMENTO, SIQUEIRA, BERTOLOTO, GASPAR, 2012).

BIBLIOGRAPHICAL REFERENCES

ALENCAR, E. M. L. S. de; FLEITH, D. de S. Barreiras a promopao da criatividade no ensino fundamental. **Psic.: Teor. e Pesq.,** Brasilia , v. 24, n. 1, p. 59-65,2008 .

ANHUSSI, E. C. The use of newspapers in the classroom: their importance and teachers' conceptions. 2009. 156 f. Diss. **Dissertation (Master's Degree in Education)**-Faculty of Philosophy and Sciences, Universidade Estadual Paulista, Presidente Prudente, 2009.

ARAUJO, S. A. de. Pedagogical possibilities of cinema in the classroom. **Revista Espapo Academico** - n° 79, p.1-4, 2007.

BRANDAO, I.D.N.; MELLO, M. C. O. Didactic resources in the teaching of Geography: thematizations and possibilities of use in pedagogical practices. **Geografia e Pesquisa** (UNESP. Ourinhos), v. 7, p. 8197, 2013.

BRITO, R.B. de; FREIRE, E.C.S . The Seventh Art in Education: Cinema as an Educommunicative Laptop. IN: **XV Encontro Latino Americano de Iniciapao Cientifica e XI Encontro Latino Americano de Pos-Graduapao** - Universidade do Vale do Paraiba, p.1-5,2011.

CAMPOS, R. R. de. Cinema, Geography and the classroom. **Estudos Geograficos**, Rio Claro, v°4, n°1, p.1-22, 2006.

CAPELETTO, A. **Biologia e Educapao ambiental: Roteiros de trabalho.** Editora Atica, 1992. p. 224.

CASTELLAR, S. M. V. Teacher training and the teaching of geography. **Terra Livre**, Sao Paulo, n. 14, p. 51-59, 1999.

CASTELLAR, S; VILHENA, J. **Ensino de geografia.** Sao Paulo: CENGAGE Learning, Colepao ideias em apao, 161p,2010.

CAVALCANTI, L. de S. Geography and contemporary school reality: advances, paths, alternatives. IN: **I Seminario Nacional: Curriculo em Movimento-Perspectivas Atuais.** Belo Horizonte. Proceedings of the Seminar, 2010, 16p. Available at:<
http://portal.mec.gov. br/docman/dezembro-2010-pdf/7167-3-3-geografia-realidade-escolar-lana-souza/file>. Accessed on: July 4, 2018.

DIAS, L. S.; MARQUES, M.D.; DIAS, L.S. Educapao, Educapao Ambiental, Perceppao Ambiental E Educomunicapao **IN: Educapao Ambiental: conceitos, metodologia e praticas**. (Orgs.): DIAS ,L.S,LEAL, A.C; JUNIOR, S.C. ANAP, Tupa, Sao Paulo, 1ª ed;187p,2016.

FERREIRA, A.A.; RODRIGUES, S.X.V.; JESUS, J.N de. THE IMPORTANCE OF TEACHING PRACTICE IN GEOGRAPHY. IN: **IV EDIPE - Encontro Estadual de Didatica e Pratica de Ensino**, Universidade Estadual de Goias (UEG) .p.1-10.2011.

FRANQA, B. A. The use of didactic resources in geography classes in schools in the western zone of Rio de Janeiro. IN: **10th National Meeting of Teaching Practice in Geography (ENPEG)**. Porto Alegre, 2009, 8p. Available at:<
http://www.agb.org.br/XENPEG/artigos/Poster/P%20(6).pdf>. Accessed on: August 5, 2018.

GONQALVES, F. do. C.; FLEITH, D. de. S.; LIBORIO, A.C.O. Creativity in class: perception of students from two Brazilian states. **Arquivos Brasileiros de Psicologia**, v. 63, n. 1, 2011.

Junior, A. M. A., Nascimento, L. P., Siqueira, S. A., Bertoloto, J. C., Gaspar, B. F. L. A Produpoo De Material Didatico-Pedagogico Em Geografia Para O Ensino Fundamental: notas de uma experiência. **PerCursos,** v. 13, n. 2, p. 75-93, 2012.

KAERCHER, N.A. Know and reveal yourself by studying the city. **Geografia: Praticas Pedagogicas** Vol. 2, p. 121, 2009.

MALANSKI, L. M. **The newspaper as a teaching resource for pedagogical practices in geography**. 2010. Available at: <http://pt.scribd.com/doc/53401447/O-jornalcomo-recurso-didatico- para-as-praticas-pedagogicas-em-Geografia>. Accessed on: Aug. 12, 2014.

OLIVEIRA, E.B.P.; ALENCAR, E.M.L.S.de. Importance of creativity in school and in teaching work according to pedagogical coordinators. **Estudos de Psicologia,** v. 29, n. 4, p. 541-552, 2012.

PAULA, C. P. de.; TORRES, E. C. **O Uso** De **Jornal Como Instrumento Pedagogico No Ensino De Geografia**. Available at: <http://www.diaadiaeducacao.pr.gov.br/portals/cadernospde/pdebus ca/producoes_pde/2014/2014_uel_geo_artigo_claudia_ponciano_d e_paula.pdf>. Accessed on: August 6, 2018.

PERUSI, M.C.; SENA, C.C.R.G de. Soil Education, Inclusive Environmental Education and Continuing Teacher Education: Multiple Aspects of Geographical Knowledge. **Revista Entre-Lugar**, Dourados, MS, ano 3, n.6, ed. 2. p.153-164, 2012.

PIAGET, J. **The formation of symbols in children**. Rio de Janeiro, RJ: Zahar, 1975.

ROSSINI, M. de. S. RENNER, A.G. New visual culture? Netflix and the changing process of audiovisual production, distribution and consumption. IN: **XXXVIII Brazilian Congress of Communication Sciences**, Rio de Janeiro, p.1-13, 2015.

SANTOS, F, K, S. Supervised Curricular Internship in Geography and Pedagogical Mediation: between knowledge and practices for a dialogical posture. **Revista Brasileira de Educagao Geografica**, Campinas, v.4, n.7, p 85-99, 2014.

SILVA, B. N.S. Cinema and the classroom: a path for training. **Revista espago academica**, 2009.

SILVA, E.T. **A produgao da leitura na escola: pesquisas e propostas.** 5ª ed. Sao Paulo: Atica, 2005.

SILVA, L.C da; BERTAZZO, C.J. O ludico, a geografia e a mediação didatica. **Revista Geoaraguaia,** v. 3, n. 2, 2013.

SOUZA, C. J. B de.; SILVA, B. A. da.; OLIVEIRA, E. S. de.; RODRIGUES, T.C. Approach to Soils in Science Teaching Through Differentiated Practical Activities. IN: **III Congresso Nacional de Educagao** - CONEDU, p.2-6, 2016.

THIESEN, J. da S. Geografia escolar: dos conceitos essenciais às formas de abordagem no ensino. **Geography Teaching and Research.** V. 15, n. 1, 2011, 85-95p.

Buy your books fast and straightforward online - at one of world's fastest growing online book stores! Environmentally sound due to Print-on-Demand technologies.

Buy your books online at
www.morebooks.shop

Kaufen Sie Ihre Bücher schnell und unkompliziert online – auf einer der am schnellsten wachsenden Buchhandelsplattformen weltweit! Dank Print-On-Demand umwelt- und ressourcenschonend produziert.

Bücher schneller online kaufen
www.morebooks.shop

info@omniscriptum.com
www.omniscriptum.com

Printed by Books on Demand GmbH, Norderstedt / Germany